T0177432

Protocols for Pre-Field Screening of Mutants for Salt Tolerance in Rice, Wheat and Barley

Souleymane Bado • Brian P. Forster •
Abdelbagi M.A. Ghanim •
Joanna Jankowicz-Cieslak • Günter Berthold •
Liu Luxiang

Protocols for Pre-Field Screening of Mutants for Salt Tolerance in Rice, Wheat and Barley

Souleymane Bado
Plant Breeding and Genetics Laboratory
Joint FAO/IAEA Division of Nuclear
   Techniques in Food and Agriculture
Vienna, Austria

Brian P. Forster
Plant Breeding and Genetics Laboratory
Joint FAO/IAEA Division of Nuclear
   Techniques in Food and Agriculture
Vienna, Austria

Abdelbagi M.A. Ghanim
Plant Breeding and Genetics Laboratory
Joint FAO/IAEA Division of Nuclear
   Techniques in Food and Agriculture
Vienna, Austria

Joanna Jankowicz-Cieslak
Plant Breeding and Genetics Laboratory
Joint FAO/IAEA Division of Nuclear
   Techniques in Food and Agriculture
Vienna, Austria

Günter Berthold
Plant Breeding and Genetics Laboratory
Joint FAO/IAEA Division of Nuclear
   Techniques in Food and Agriculture
Vienna, Austria

Liu Luxiang
Chinese Academy of Agricultural Sciences
Beijing, China

ISBN 978-3-319-26588-9   ISBN 978-3-319-26590-2   (eBook)
DOI 10.1007/978-3-319-26590-2

Springer Cham Heidelberg New York Dordrecht London
© International Atomic Energy Agency 2016. The book is published with open access at SpringerLink.
com. Open Access provided with a grant from the International Atomic Energy Agency

Printed on acid-free paper

Springer International Publishing AG Switzerland is part of Springer Science+Business Media
(www.springer.com)

Preface

The Joint FAO/IAEA Programme of Nuclear Techniques in Food and Agriculture has supported Member States in the use of nuclear techniques in plant breeding and genetics for over 50 years. This has been achieved through research and training especially in developing methods for mutation induction and mutation detection. Mutation induction in plants aims to generate novel genetic diversity for plant breeders targeting yield, quality, resistance to pests and diseases, and tolerance to abiotic stresses such as salinity. Induced mutation in plants began in the 1920s and the first mutant cultivar was "Vorstenland" tobacco released in Indonesia in 1934. Plant mutation breeding has been very successful, and today, there are over 3220 officially released mutant cultivars in over 210 crop species worldwide. World food security continues to be threatened notably by climate change, lack of agricultural land, and a growing human population. Thus, there is continual pressure on plant breeders to develop higher yielding crop cultivars. Plant mutation breeding can help meet these demands. One issue, however, is the ability to select mutants carrying desired traits as this requires the development of screening protocols. This booklet provides a simple protocol to screen for mutants in cereal crops tolerant to salinity.

The booklet has three main sections: (1) a brief introduction to the problem of soil salinity, (2) a protocol for measuring soil salinity, and (3) a protocol for screening for salt-tolerant cereal genotypes. The protocols are aimed to assist plant breeders and especially breeders who need to screen cereal populations, such as mutant populations, for salt tolerance. The protocols are designed to be effective, low cost, and user friendly.

The booklet provides simple and quick methods for soil sampling and analysis for water-soluble salt content, both of which are critical for the downstream screening. With these easy-to-follow protocols, users can conduct analyses in a quick and effective manner.

Simple and quick methods are also provided to screen seedlings for salt tolerance in hydroponics. The seedling test takes 4–6 weeks and allows the screening of several hundred seedlings. The test can be used to screen segregating populations, standard lines and cultivars, as well as M_2 populations and advanced mutant generations. A list of the required equipment is given, along with setup procedures for hydroponics hardware and stock solutions. Tolerance is determined by performance comparisons against known salt-tolerant genotypes. Control tests (without salt) can also be performed if required as an indicator that the system is working and for comparing growth under salt and non-salt conditions. Indicators of tolerance are leaf colour, leaf rolling, leaf tip dying, and seedling death. Root damage (growth and browning) and biomass can also be observed. The protocol was originally designed to screen rice mutant populations for salinity, but has been adapted for wheat and barley by the addition of aeration and increased salt concentrations. The protocol has been tested and validated on materials from Iran, Myanmar, and Vietnam.

Vienna, Austria

Brian P. Forster

Acknowledgements

First at all, the authors would like to thank Dr. Afza Rownak for the knowledge she shared on mutation induction and screening for stress tolerance. We would also like to thank the following for their evaluation and useful comments to improve this protocol: Dr. Ping An, Arid Land Research Center, Tottori University, Japan, Prof. Kamal El-Siddig, Agricultural Research Corporation, Ministry of Agriculture, Republic of Sudan, Dr. Nina Nurlina, Hasanuddin University, Indonesia, and Dr. Mohammad A.K. Azad, Bangladesh Institute of Nuclear Agriculture, Bangladesh.

Contents

Chapter 1
Introduction

Abstract Salinity is a major abiotic stress limiting crop yields in many parts of the world. The FAO (Food and Agriculture Organization) Land and Plant Nutrition Management service estimates that over 400 million hectares (6 %) of the Earth's land is affected by salt. Breeding for salt tolerance is a major goal for cereal researchers for which screens are required to select out tolerant lines. Screening for salt tolerance in the field is difficult as soil salinity is dynamic, the level of salt varies both horizontally and vertically in the soil profile and changes with time. These environmental perturbations can be overcome by testing in hydroponic system where the testing environment is controlled.

1.1 Background

Soil salinity affects more than 800 million hectares worldwide, equivalent to over 6 % of all land on Earth. Of the 1500 million hectares cultivated in dry regions, 2 % are affected by salt. Of the 230 million hectares that are irrigated, 20 % are salt affected (Munns 2005). Irrigation exacerbates the problem as the irrigation waters bring dissolved salts which are deposited in the soil. History tells us of several civilisations collapsed because of salinisation of agricultural land due to irrigation, for example, the ancient Mesopotamian civilisation (now part of Iraq) faded away some 4400–3700 years ago due to crop failures caused by salinity. Crop records of Sumeria indicate a change of crop from wheat (salt sensitive) to barley (salt tolerant) and then a subsequent decline of barley yields as soils became increasingly saline. The Peruvian culture of the Viru Valley, which peaked 1200 years ago, was forced to retreat up into the highlands because of salinisation of fields (Pearce 1987; Jacobsen and Adams 1968).

Irrigation without adequate soil and salt management systems inevitably leads to salinisation of cultivated land. This is due to continual additions of soluble salts of sodium, calcium, magnesium and potassium, usually as chlorides or sulphates, which are concentrated in the soil as water is lost due to evaporation and crop plant transpiration. In addition, excess sodium (sodicity) promotes slaking of soil

© International Atomic Energy Agency 2016
S. Bado et al., *Protocols for Pre-Field Screening of Mutants for Salt Tolerance in Rice, Wheat and Barley*, DOI 10.1007/978-3-319-26590-2_1

aggregates that degrades the soil structure and impedes water movement and root growth.

Saline environments are generally grouped as being either wet or dry. Wet saline habitats tend to occur in coastal regions and are dominated by salt marshes. Since these areas border the sea, they are subject to periodic inundations, and as a result the level of salinity fluctuates over time. Dry saline habitats are usually located inland, often bordering deserts (Tal 1985; Neumann 1997; Flowers 2004). Other types of saline environments include seashore dunes, where salt spray is a salinising factor, and dry salt lakes. Common features of saline environments are the salinity of the soil and/or of their associated water resources and specialised flora and fauna. The most abundant salts in saline soils are sodium chloride (NaCl) and sodium sulphate (Na_2SO_4), which may be associated with magnesium (Mg) salts.

Sustainable irrigation systems incorporate one or more forms of leaching and drainage of brackish water (slightly saline water). Leaching may be achieved by natural rainfall and run-off or by irrigation with fresh water and artificial drainage systems. In both systems, drainage needs to be provided. These may be small scale for subsistence farming communities or may involve massive civil engineering projects such as the West Bank Outfall drain of the river Indus in Pakistan (Khan et al. 2013). Cropping systems also need to be devised that maximise the benefit of seasonal conditions, e.g. exploitation of monsoon rains to leach out salts and early maturing crops that avoid high saline periods.

With increasing human populations, there is an increasing demand for food. Throughout the world, the best agricultural land is already fully utilised, and hence marginal land, including saline land, is being brought into agriculture. Unfortunately, most crop plants are sensitive to salt (glycophytes). Salinity is therefore a major environmental constraint to crop production throughout the world.

1.2 Biology of Salt Tolerance

Salt-tolerant plants have evolved in many taxa of the plant kingdom. Aronson (1989) noted over 100 plant families which contain salt-tolerant species. Most plant families contain a few salt-tolerant species (halophytes), but the Chenopodiaceae is an exception in containing over 350. It has been suggested that salt tolerance evolved in many higher plants as a consequence of becoming established in estuaries (O'Leary and Glenn 1994) and then spreading to inland environments. More than 30 % of extant plant families have halophytic members (*circa* 2500 species) which are mainly found in salt marshes or desert flats (Glenn 1997).

Ungar (1991) defined salt-loving plant, halophytes, as those that tolerate relatively high soil salinity and are capable of accumulating relatively high quantities of sodium and chloride; glycophytes on the other hand are defined as species that show little tolerance to elevated saline levels in the root zone and do not accumulate high concentrations of salts in growing tissues and organs. Extreme halophytes such as

Salicornia europaea and *Suaeda maritima* can tolerate saline water above that of sea water, whereas glycophytes are intolerant of salinities above 10 % of sea water.

In general, three physiological mechanisms are deployed by plants growing in saline conditions: (1) osmotic adjustment, (2) ion exclusion and (3) tissue tolerance to accumulated ions. The effects of salinity are first observed by a reduction in plant growth (Munns 1993), which has two response phases: (1) a rapid response to the increase in external osmotic pressure (the osmotic phase), which starts as soon as the salt concentration increases around the roots to a threshold level (approximately 40 mM NaCl for most plant), and (2) a slower response in which harmful ions accumulate in leaves (the ionic phase). When the death rate of older leaves is greater than the production of new leaves, the photosynthetic capacity will no longer be optimum and growth rate retards (Munns and Tester 2008).

Genetic variation exists for these major mechanisms of salt stress (osmotic stress, ion exclusion and tissue tolerance) and their component parts (ion compartmentalisation, ion transport, toxicity, etc.). Genetic variation can be found within as well as between species. The former is good news for plant breeders as it allows salt tolerance traits to be transferred through normal cross-breeding, whereas interspecific crosses may provide a means of transferring genes from one species (a donor) to another (a recipient).

1.3   Screening Methods

Plant growth responses to salinity vary with plant life cycle; critical stages sensitive to salinity are germination, seedling establishment and flowering (Ashraf and Waheed 1990; Flowers 2004). Criteria for evaluating and screening salinity tolerance in crop plants vary depending on the level and duration of salt stress and the plant developmental stage (Shannon 1985; Neumann 1997). In general, tolerance to salt stress is assessed in terms of biomass production or yield compared to non-stress conditions. In conditions of low to moderate salinity, the production capacity of the genotype is often the most pertinent measure, whereas survival ability is often used at relatively high salinity levels (Epstein et al. 1980). The physiological mechanisms that play a major role in maintaining the production capacity of a genotype are not the same as those that contribute to tolerance at extremely high salt concentrations.

Genotypes are generally evaluated using phenotypic observations.

Phenotypic selection parameters include:

(a)   Germination

Germination tests are easy to perform and may be important where the crops are required to germinate and establish in saline conditions. However, germination in saline conditions is not often associated with salinity tolerance in subsequent growth stages (Dewy 1962; Shannon 1985; Flowers 2004).

(b) Plant survival

Selection on the basis of plant survival at high salt concentrations has been proposed as a selection criterion for tomato, barley and wheat (Rush and Epstein 1976; Espstein and Norlyn 1977). The ability of a genotype to survive and complete its life cycle at very high salinities, irrespective of yield potential under moderate salinity levels, is considered as being tolerant in the absolute sense.

(c) Leaf damage

Since most crops are glycophytes, they are unable to restrict toxic salt ions being translocated from roots into shoots and leaves. Consequently, salinity damage may be readily observed by leaf symptoms of bleaching and necrosis. Screening for salt tolerance by leaf damage is therefore common (Richards et al. 1987; Gregorio et al. 1997).

(d) Biomass and yield

For plant breeders, yield and biomass are obvious parameters in assessing salt tolerance (Richards et al. 1987). These parameters, however, do not provide information on the underlying physiological mechanisms. In the past, plant breeders have not been interested in physiological mechanism; that a genotype was tolerant was sufficient, the physiological mechanisms were regarded as academic. However, with the emergence of gene function studies, this view is changing.

(e) Physiological mechanisms

Physiological mechanisms that confer tolerance to salt may be harnessed for screening. These may include measurements of tissue sodium content, ion discrimination and osmotic adjustment. Surrogates such as carbon isotope discrimination ($\delta^{13}C$) which give a general indication of plant stress may also be used (Flowers and Yeo 1981; Pakniyat et al. 1997).

1.4 Breeding for Salt Tolerance

1.4.1 Traditional Breeding

Subbarao and Johansen (1994) suggested the following pragmatic considerations in initiating a programme for genetic improvement of crop plants:

1. Define the target environment.
2. Define the level of improvement necessary.
3. Define the growth stage response.
4. Choose the screening method.
5. Choose the selection criteria.
6. Assess the genotypic variation for the various traits under consideration that may have a functional role in improving salinity tolerance.
7. Identify genetic resources for the various components (traits) of salinity tolerance.

8. Determine the genetic basis for traits under consideration, and estimate their heritability.
9. Initiate breeding programmes that combine various traits from different sources into a locally adapted germplasm for ultimate development of a salt-tolerant cultivar.
10. Test selected genotypes in target locations, in a range of saline soils within a production environment, to assess their potential adaptability as new cultivars.

As a comparison, Flowers and Yeo (1995) suggested five strategies in developing salt-tolerant crops:

1. Develop naturally tolerant species (halophytes) as alternative crops.
2. Use interspecific hybridisation to raise the tolerance of current crops.
3. Exploit genetic variation already present in crop gene pools.
4. Generate variation within existing crops by using recurrent selection, mutation induction and/or tissue culture.
5. Breed for yield rather than tolerance.

The use of conventional cross-breeding for salt tolerance has met with little success. This is largely because the required salt tolerance is not present in the primary gene pool of breeding materials. For many crop species, salt tolerance traits are not present in the secondary gene pool (within the species), and for some crops breeders have to resort to interspecific and intergeneric crosses involving wild species to tap into genes that may be transferred by sexual reproduction and recombination. As a consequence, novel genetic variation needs to be produced.

1.4.2   Induced Mutation in Breeding for Salt Tolerance

Mutation induction is one means of increasing biodiversity in crop plants. Mutation induction can be achieved within minutes by gamma-ray or X-ray irradiation of plant materials (usually seed). Mutation may also be produced easily through the use of chemical agents. The detection of mutants carrying the desired variation is more time-consuming and usually involves the screening of thousands of individuals either phenotypically (response to salinity) or genotypically (searching for changes in target genes). Screening for desired mutants is often a major bottleneck in crop improvement.

Once desired mutants are found, these may be entered directly into breeding programmes. However, it is more common that some pre-breeding is performed to 'clean up' the genetic background of the mutant lines before entry into breeding programmes. Various genetic marker techniques may be deployed in marker-assisted selection to increase breeding efficiency.

1.5   Need for Reliable Screening Techniques for Pre-field Selection

Salt-affected soils can be classified into three types: (1) saline, (2) sodic and (3) saline–sodic soils. Soil salinity can decrease water availability in the soil and produce toxic effects on particular plant processes. Measuring soil salinity is difficult as it varies with space and time. As a result, soil must be sampled at various times in various places to analyse the effects of salinity on plant growth. A large number of samples are needed to characterise a specific field fully, and sampling should follow all changes in conditions; thus, in many cases, soil sampling requires considerable time and effort in the field.

Abiotic stress tolerance, especially salinity stress, is complex because of variation in sensitivity at various stages in the life cycle. Rice is comparatively tolerant to salt stress during germination, active tillering (vegetative growth) and the later stages of maturity. It is most sensitive during seedling establishment and reproductive stages. Screening at an early growth stage (2–4 weeks) is more convenient than at flowering. This is due to the fact that it is (1) quick, (2) seedlings take up less space, (3) tolerant seedlings may be recovered for seed production and (4) seedling tests are more efficient in terms of time and costs. Seedling screening offers the possibility of preselection of putative individual mutants, mutant populations, breeding lines and progeny and cultivars before large-scale field evaluation.

The rice seedling test described in this booklet is an adaptation of that originally devised in collaboration with the International Rice Research Institute (IRRI). The current system however does not use a floating support and is designed to be robust, reusable and multiple functional; it can be adapted to evaluate individual genotypes or large mutant populations. The hydroponics set-up uses plastic tanks with tight-fitting polyvinyl chloride (PVC) support plates (platforms). A prototype system used bulky styrofoam supports, but these are difficult to maintain and become brittle and contaminated with algae and other microbes with time. The PVC supports are more robust, easily cleaned and can be used repeatedly with minimal maintenance. The PVC platforms are also strong enough to support several hundred seedlings. The test is rapid taking 4–6 weeks. A simplified non-aerated system is used for rice, but forced air aeration and higher salt concentrations are used in screening wheat and barley seedlings.

References

Aronson J (1989) HALOPH a data base of salt tolerant plant of the world. Office of Arid Land Studies, University of Arizona, Tuscon

Ashraf M, Waheed A (1990) Screening of local/exotic accessions of lentil (*Lens culinaris* Medic) for salt tolerance at two growth stages. Plant and Soil 128:167–176

Dewy DR (1962) Breeding crested wheatgrass for salt tolerance. Crop Sci 2:403–407

Epstein E, Norlyn JD, Rush DW, Kingsbury RW, Kelly DB, Cunningham GA, Wrona AF (1980) Saline culture of crops: a genetic approach. Science 210:399–404

Epstein E, Norlyn JD (1977) Seawater-based crop production: a feasibility study. Science 197:247–261

Flowers TJ (2004) Improving crop salt tolerance. J Exp Bot 55:307–319

Flowers TJ, Yeo AR (1981) Variability in the resistance of sodium chloride salinity within rice [*Oryza sativa* L.] varieties. New Phytol 88:363–373

Flowers TJ, Yeo AR (1995) Breeding for salinity resistance in crop plants—where next. Aust J Plant Physiol 22:875–884

Glenn EP (1997) Mechanisms of salt tolerance in higher plants. In: Basra AS, Basra RK (eds) Mechanisms of environmental stress resistance in plants. Harwood Academic Publishers, Amsterdam, pp 83–110

Gregorio GB, Senadhira D, Mendoza RT (1997) Screening rice for salinity tolerance, vol 22, IRRI discussion paper series. IRRI, Manila, p 30

Jacobsen T, Adams RM (1968) Salt and silt in ancient Mesopotamian agriculture. Science 128:1251–1258

Khan MZ, Jabeen T, Ghalib SA, Siddiqui S, Alvi SM, Khan IS, Yasmeen G, Zehra A, Tabbassum F, Sharmeen R (2013) Effect of right bank outfall drain (RBOD) on biodiversity of the wetlands of Haleji wetland complex, Sindh. SCRO Res Annu Rep 1:48–75

Munns R (1993) Physiological processes limiting plant-growth in saline soils: some dogmas and hypotheses. Plant Cell Environ 16:15–24

Munns R (2005) Genes and salt tolerance: bringing them together. New Phytol 167:645–663

Munns R, Tester M (2008) Mechanisms of salinity tolerance. Annu Rev Plant Biol 59:651–681

Neumann P (1997) Salinity resistance and plant growth revisited. Plant Cell Environ 20:1193–1198

O'Leary J, Glenn E (1994) Global distribution and potential for halophytes. In: Squires VR, Ayoub AT (eds) Halophytes as resource for livestock and for rehabilitation of degraded lands. Kluwer Academic Publishers, Dordecht, pp 7–17

Pakniyat H, Handley LL, Thomas WTB, Connolly T, Macaulay M, Caligari PDS, Forster BP (1997) Comparison of shoot dry weight, Na^+ content and $\delta13C$ values of *ari-e* and other semi-dwarf barley mutants under salt stress. Euphytica 94:7–14

Pearce F (1987) Banishing the salt of the earth. New Sci 11:53–56

Richards RA, Dennett CW, Qualset CO, Epstein E, Norlyn JD, Winslow MD (1987) Variation in yield of grain and biomass in wheat, barley and triticale in a salt-affected field. Field Crops Res 15:277–278

Rush DW, Epstein E (1976) Genotypic response to salinity: differences between salt sensitive and salt tolerant genotypes in the tomato. Plant Physiol 57:162–166

Shannon MC (1985) Principles and strategies in breeding for higher salt tolerance. Plant Soil 89:227–241

Subbarao GV, Johansen C (1994) Potential for genetic improvement in salinity tolerance in legumes: pigeon pea. In: Pessarakli M (ed) Handbook of plants and crop stress. Dekker, New York, pp 581–595

Tal M (1985) Genetics of salt tolerance in higher plants: theoretical and practical considerations. Plant Soil 89:199–226

Ungar IA (1991) Ecophysiology of vascular halophytes. CRC Press, Boca Raton

Chapter 2
Objectives

Abstract Salinity affects soil, water and crop plants. The severity of soil salinity needs to be determined in order to make informed decisions on best cropping practices. Likewise, the tolerance of crop cultivars needs to be matched to the growing conditions. Protocols are therefore required to monitor field salinity and to evaluate crop tolerance to salt.

2.1 Monitoring Field Salinity

Soil salinity affects both, water availability and plant growth processes. Salinity refers to the presence of one or more of a number of dissolved inorganic ions (Na^+, Mg^{2+}, Ca^{2+}, K^+, SO_4^{2-}, HCO_3^-, NO_3^- and CO_3^{2-}) in the soil. Monitoring of soil salinity and the preparation of soil salinity maps are essential objectives for good management of salt-affected lands and the productive agriculture of salt-tolerant crop cultivars.

2.2 Screening for Salt Tolerance

The aims are to provide a screen in which salt-tolerant rice, wheat and barley lines can be selected for use in plant breeding. The screen may also be used to compare and classify salt tolerance in a range of germplasm. Extensive tests have been carried out at the IAEA's Plant Breeding and Genetics Laboratory (PBGL) using rice genotypes with known susceptibility/tolerance to saline field conditions. Correlations have been established between seedling hydroponics responses and field salinity tolerance. Thus, the seedling screen described here can be used to select plants that may be expected to perform well in saline field conditions.

© International Atomic Energy Agency 2016
S. Bado et al., *Protocols for Pre-Field Screening of Mutants for Salt Tolerance in Rice, Wheat and Barley*, DOI 10.1007/978-3-319-26590-2_2

2.3   Benefits and Drawbacks of Seedling Screening

The protocols described in this book use seedlings as the test materials. Tolerance to salt at the seedling stage has been correlated with field performance (Zeng et al. 2003) and in the test cases given in Chap. 4 (Tables 4.6, 4.7 and 4.8), and selection on the basis of plant survival at high salt concentrations has been proposed as a selection criterion for several crop species (Rush and Epstein 1976; Epstein and Norlyn 1977). However, seedling screening should be regarded as a prescreen, and candidate lines should always be validated by performance in saline field conditions. Flowering time is often considered as a salt-susceptible stage and is not considered in these protocols. However, the hydroponic system may be adapted to test plants throughout their life cycling including flowering and maturity stages. The benefits and drawbacks of hydroponics screening for salt tolerance are listed in Table 2.1.

Seedling tests are best performed on M_3 or advanced populations. Tests may be done on M_2 populations which have the advantage of having relatively small population sizes, but there is a risk that the rare mutant line possessing salt tolerance is lost because of other factors, e.g. accidental miss-handling. M_3 populations and above provide more rigour as there is a degree of replication for genotypes carrying the same mutant trait.

The salt tolerance tests described in this booklet are simple and monitor seedling responses; they do not involve deep physiological understanding of the physiological mechanisms involved. Physiological aspects of salt tolerance are covered in the following references:

- Ashraf and Waheed (1990), Dewy (1962), Flowers (2004), Shannon (1985)—plant growth responses over the plant life cycle (germination to maturity)
- Shannon (1985), Neumann (1997), Epstein et al. (1980)—criteria for measuring salt stress
- Parida and Das (2005), Munns and Tester (2008)—effects on plants and mechanisms of salinity tolerance

Table 2.1 Benefits and drawbacks

Advantages	Drawbacks
• Cheap, fast and simple • Clear classification into susceptible, moderate and tolerant types • Tolerant seedlings may be recovered • High-throughput screen • Preselection technique for putative mutants • Equipment is reusable • Greater uniformity compared to soil-based salt tolerance screening	• Requires continual vigilance and maintenance (replenishment of test solution every 2 days) • Solutions need to be changed; therefore, adequate stocks of chemicals are required • Requires good-quality growing conditions • Homogenous, good seed quality required

References

Ashraf M, Waheed A (1990) Screening of local/exotic accessions of lentil (*Lens culinaris* Medic) for salt tolerance at two growth stages. Plant Soil 128:167–176

Dewy DR (1962) Breeding crested wheatgrass for salt tolerance. Crop Sci 2:403–407

Epstein E, Norlyn JD (1977) Seawater-based crop production: a feasibility study. Science 197:247–261

Epstein E, Norlyn JD, Rush DW, Kingsbury RW, Kelly DB, Cunningham GA, Wrona AF (1980) Saline culture of crops: a genetic approach. Science 210:399–404

Flowers TJ (2004) Improving crop salt tolerance. J Exp Bot 55:307–319

Munns R, Tester M (2008) Mechanisms of salinity tolerance. Annu Rev Plant Biol 59:651–681

Neumann P (1997) Salinity resistance and plant growth revisited. Plant Cell Environ 20:1193–1198

Parida AK, Das AB (2005) Salt tolerance and salinity effects on plants: a review. Ecotoxicol Environ Saf 60:324–349

Rush DW, Epstein E (1976) Genotypic response to salinity: differences between salt sensitive and salt tolerant genotypes in the tomato. Plant Physiol 57:162–166

Shannon MC (1985) Principles and strategies in breeding for higher salt tolerance. Plant Soil 89:227–241

Zeng L, Poss JA, Wilson C, Draz AE, Gregorio GB, Grieve CM (2003) Evaluation of salt tolerance in rice genotypes by physiological characters. Euphytica 129:281–292

Chapter 3
Protocol for Measuring Soil Salinity

Abstract A simple protocol is described that tests soil salinity. Water-soluble salts are extracted from soil samples and salt content measured. Accurate field evaluations require sampling at various field locations and various depths and over time take into account the crop species to be grown. Instruments and reagents are listed in preparing soil–water extracts and for measuring salt content. Two methods are provided in measuring salt content, by weight and by electrical conductivity.

3.1 Background

The measurement of soil salt content is very important for plant salt tolerance studies. The most commonly used method is a simple field test. The characteristics of saline soil areas include microtopography, complicated soil types and significant differences in local soil conditions. In order to reduce testing errors caused by differences in local soil conditions, numerous samples are required and repeat sampling needs to be performed. The soil samples should be collected from different soil layers at different depths based on the plant species root growth. For deep-rooted plants, sample soil layers from 0 to 5 cm, 5 to 10 cm, 10 to 20 cm, 20 to 40 cm, 40 to 60 cm and so on are required to a depth of at least 1 m. The samples from different layers should be mixed uniformly. For plants with shallow rooting systems, soil layers should be sampled to a depth of about 60 cm. The salt content in saline soils is dynamic and changes over time and is heterogeneous from location to location. It also changes with the year and the month and even during a day. Considering seasonal and climatic conditions, the sampling times should include spring and summer when salts tend to accumulate and autumn and winter when rains tend to leach salts from the soil. The growing season, time of sowing, seedling establishment, flowering time and harvest should also be considered as some growth stages may be more sensitive to salt damage than others, depending on the crop. Data collected over the years is also useful in assessing trends in salinisation.

Saline soils possess excessive water-soluble salts. Measuring water-soluble salts has two main steps: (1) the preparation of the sample solution according to the

© International Atomic Energy Agency 2016

S. Bado et al., *Protocols for Pre-Field Screening of Mutants for Salt Tolerance in Rice, Wheat and Barley*, DOI 10.1007/978-3-319-26590-2_3

specific water/soil ratio and (2) the analysis of the soil salt concentration and ionic components in the soil sample. In general, studies on dynamic changes of water and salt content in the soil use a water/soil ratio of 5:1, whereas a water/soil ratio of 1:1 is generally used for the analysis of alkaline soils. The method of saturated soil extract is rarely used because the execution of this method is tedious, and it is difficult to determine the correct saturation point. The sample solution in the following tests refers to 5:1 water/soil extract.

3.2 Instruments and Reagents

Instruments: reciprocating shaker, 1/100 balance, Buchner funnel, vacuum pump, centrifuge (4000 r/min), gas extraction bottle.

Reagent: 0.1 % $NaPO_3$.

3.3 Preparation of 5:1 Water/Soil Extract

Weigh 100 g air-dried soil sampled from the field that passes through a 1 mm sieve. Put the soil sample in an Erlenmeyer flask. Add 500 ml CO_2-free distilled water (based on a water/soil ratio of 5:1). Seal the flask mouth with a rubber stopper and place the Erlenmeyer flask in a reciprocating shaker and shake for 3 min. Immediately, after shaking, perform an air pump filtration with a Buchner funnel. Collect the clear liquid in a 500 ml Erlenmeyer flask. Add 1 drop of 0.1 % $NaPO_3$ for every 25 ml.

3.4 Preparation of 1:1 Water/Soil Extract

Weigh an air-dried soil sample that passes through a 1 mm sieve. Put the soil sample in an Erlenmeyer flask. Add CO_2-free distilled water based on a water/soil ratio of 1:1. The rest of the operation is the same as above.

3.5 Important Considerations

• When extracting with the 5:1 water/soil ratio, hygroscopic water of the air-dried soil can be ignored due to the high percentage of water. When extracting with the 1:1 water/soil ratio, hygroscopic water of the air-dried soil must be corrected to avoid test error (compared to the 5:1 water/soil ratio, hygroscopic water in the soil may affect the water/soil ratio at 1:1; this needs to be corrected to avoid test errors, and therefore the use of completely air-dried soil is recommended).

- In the process of extraction of water-soluble salts in soil, 3 min shaking is sufficient for the water-soluble chlorides, carbonates and sulphates to dissolve in the water. With an extended shaking time or standing time, the neutral salts and water-insoluble salts will also enter the extract and cause greater errors.
- Both the partial pressure of CO_2 in the air and the dissolved CO_2 in the distilled water will affect the solubility of some salts including $CaCO_3$, $CaSO_4$ and $MgSO_4$. As a result, the salt content in the extract will be affected. Therefore, CO_2-free distilled water must be used in the extraction.
- The standing time of the sample solution should not exceed 1 day.
- Adding a small amount of $NaPO_3$ solution in the soil extract can prevent the formation of a $CaCO_3$ precipitate when standing. Although $NaPO_3$ will slightly increase the Na^+ concentration in the extract, the error caused by $NaPO_3$ is much smaller than the error caused by $CaCO_3$ precipitate.

3.6 Measurements

The main methods of measuring total water-soluble salts in a soil sample are the (1) weight method and (2) conductivity method. The data obtained from the weight method are reliable, but the operation is tedious and time-consuming. The conductivity method is simple.

3.6.1 Weight Method

This method is based on a water extract from a soil sample. The extract is evaporated to dryness and then dried at 105–110 °C to constant weight. The total dried residue contains both water-soluble salts and water-soluble organic matter. H_2O_2 is used to remove the organic matter in the residue. What remains are the total water-soluble salts from the soil.

3.6.1.1 Instruments and Reagents

Instruments: evaporating dish, water bath, dryer, electrothermal drying oven, analytical balance.
    Reagents: 15 % H_2O_2 and 2 % Na_2CO_3.

3.6.1.2 Method

Draw 50.0 ml of solution from a soil sample of known weight (w), place in an evaporating dish and weigh (w_0). Evaporate to dryness in a water bath and then dry in an electrothermal drying oven at 105–110 °C for 4 h. Remove from the oven and place in a dryer for 30 min, then weigh using an analytical balance. Return sample to the electrothermal drying oven for 2 more hours, cool down and reweigh. Repeat these steps until a constant weight (w_1) is obtained; the weight difference between the two times should not be more than 1 mg. Calculate the weight of the dried residue.

Add 15 % H_2O_2 in drops to wet the residue. Evaporate to dryness in a water bath. Repeat this treatment until the entire residue turns white. Dry the white residue to constant weight (w_2) according to the method described above. Calculate the content of the total water-soluble salts in the soil.

3.6.1.3 Calculation of Total Water-Soluble Salts

$$\text{Total dried residue} = (w_1 - w_0)/w \times 100\,\%$$

where w is the weight of the soil sample (g) that the drawn extract is equivalent to.

3.6.1.4 Important Considerations

- The volume of the soil extract to draw is determined by the soil salt content. When the soil salt content is higher than 0.5 %, draw 25 ml; when the soil content is lower than 0.5 %, draw 50 or 100 ml. Make sure that the measured total salt content is 0.02–0.2 g.
- If the residue has high contents of $CaSO_4 \cdot 2H_2O$ and $MgSO_4 \cdot 7H_2O$, drying at 105–110 °C cannot completely remove the water of crystallisation in these hydrated salts. As a result, the constant weight is difficult to obtain. In such cases, the drying temperature should be increased to 180 °C.
- If there are high amounts of $CaCl_2 \cdot 6H_2O$ and $MgCl_2 \cdot 6H_2O$ in the soil, it is difficult to get a satisfactory result even with drying at 180 °C as these salts are very hydroscopic (readily absorbed water). In such cases, first add 10 ml of 2 % Na_2CO_3. This will generate NaCl, Na_2SO_4 and $MgCO_3$ salts when evaporating to dryness. The amount of Na_2CO_3 added should then be deducted from the result of the total salt calculation.
- Since many salts absorb water from the air, the conditions for cooling and weighing should be the same.
- When using H_2O_2 to remove the organic matter, the residue only needs to be wet. Too much H_2O_2 will generate excessive foam as H_2O_2 decomposes the organic

matter. This may cause splashing and loss of salt. Repeated treatments with small amounts of H_2O_2 are recommended.

3.6.2   Conductivity Method

Water-soluble salts in the soil act as strong electrolytes. As a consequence, the soil solution has conductivity that can be measured. The electrical conductivity reflects the conductive capacity of the soil solution, and within a certain concentration range, the salt content in the soil is positively related to the electrical conductivity. But it cannot reflect the components of the mixed salt composition. If the ratios of the different salts in the soil solution are relatively constant, the salt concentration determined by electrical conductivity is very accurate. The conductivity method is a rapid and accurate method to measure soil salt content. The present tendency is to use the electrical conductivity to represent the total salt content in the soil directly. The SI unit of electrical conductivity is siemens per metre (S/m).

3.6.2.1   Instruments

Conductivity meter, thermometer ranging from 1 to 60 °C.

3.6.2.2   Method

Draw 20–30 ml of sample solution from a known amount of soil and place in a beaker. Adjust the conductivity meter according to the user's manual. Read the value of the electrical conductivity (mS) after the pointer is stable. Measure the temperature of the sample solution every 10 min.

3.6.2.3   Calculation

The electrical conductivity of the soil extract at 25 °C (EC_{25}) is used to reflect the soil salt content. It is calculated as follows:

$$EC_{25} = EC_t \times ft$$

where EC_{25} is the electrical conductivity of the soil extract at 25 °C, EC_t is the measured electrical conductivity of the soil extract at t °C and ft is the corrected value of electrical conductivity at t (see Table 3.1).

In addition, when the temperature of the soil extract is 17–35 °C, the electrical conductivity of the soil extract increases or decreases about 2 % for every 1 °C in the difference of the soil extract temperature and the standard temperature (25 °C).

Table 3.1 Corrected values of electrical conductivity at different testing temperatures

Temperature (°C)	Corrected value	Temperature (°C)	Corrected value	Temperature (°C)	Corrected value	Temperature (°C)	Corrected value
3.0	1.709	20.0	1.112	25.0	1.000	30.0	0.907
4.0	1.660	20.2	1.107	25.2	0.996	30.2	0.904
5.0	1.613	20.4	1.102	25.4	0.992	30.4	0.901
6.0	1.569	20.6	1.097	25.6	0.988	30.6	0.897
7.0	1.528	20.8	1.092	25.8	0.983	30.8	0.894
8.0	1.488	21.0	1.087	26.0	0.979	31.0	0.890
9.0	1.448	21.2	1.082	26.2	0.975	31.2	0.887
10.0	1.411	21.4	1.078	26.4	0.971	31.4	0.884
11.0	1.375	21.6	1.073	26.6	0.967	31.6	0.880
12.0	1.341	21.8	1.068	26.8	0.964	31.8	0.877
13.0	1.309	22.0	1.064	27.0	0.960	32.0	0.873
14.0	1.277	22.2	1.060	27.2	0.956	32.2	0.870
15.0	1.247	22.4	1.055	27.4	0.953	32.4	0.867
16.0	1.218	22.6	1.051	27.6	0.950	32.6	0.864
17.0	1.189	22.8	1.047	27.8	0.947	32.8	0.861
18.0	1.163	23.0	1.043	28.0	0.943	33.0	0.858
18.2	1.157	23.2	1.038	28.2	0.940	34.0	0.843
18.4	1.152	23.4	1.034	28.4	0.936	35.0	0.829
18.6	1.147	23.6	1.029	28.6	0.932	36.0	0.815
18.8	1.142	23.8	1.025	28.8	0.929	37.0	0.801
19.0	1.136	24.0	1.020	29.0	0.925	38.0	0.788
19.2	1.131	24.2	1.016	29.2	0.921	39.0	0.775
19.4	1.127	24.4	1.012	29.4	0.918	40.0	0.763
19.6	1.122	24.6	1.008	29.6	0.914	41.0	0.750
19.8	1.117	24.8	1.004	29.8	0.911		

The electrical conductivity of the soil extract at 25 °C can also be calculated according to the following formula when the soil extract temperature is 17–35 °C:

$$EC_{25} = EC_t \times [1 - (t - 25) \times 2\%]$$

where EC_{25} is the electrical conductivity of the soil extract at 25 °C, EC_t is the measured electrical conductivity of the soil extract at $t\,°C$ and t is the temperature of the soil extract (°C).

3.6.2.4   Important Considerations

- The measuring time for each sample should be relatively constant after the electrodes are inserted into the solution.
- The solutions used for conductivity measurements should be clear. Do not use liquid suspensions as these will damage the platinum back layer on the platinum electrode and cause test errors.
- Solutions with high conductivity should be diluted before taking measurements. Highly concentrated solutions will polarise the electrode and will decrease the sensitivity of the instrument.

Chapter 4
Protocol for Screening for Salt Tolerance in Rice

Abstract A simple protocol is presented that tests salt tolerance in rice seedlings. The method is based on a glasshouse hydroponics test in which salt is added to the nutrient hydroponic solution in which the seedlings are grown. A list of equipment is provided including hydroponic hardware and stock solutions. Advice is given on seed storage prior to use and pregermination treatments that promote even germination of test samples. Salt treatments commence after seedling establishment in hydroponics, at the 2–3 leaf stage. Information on responses of standard genotypes (tolerant, intermediate and sensitive) is given to which test seedlings are compared. Visual symptoms of salinity stress include reduced leaf area, whitish appearance of lower leaves, leaf tip death, leaf rolling and seedling death. Scoring is carried out according to the standard evaluation system developed by the International Rice Research Institute (IRRI). Recommended test salt concentrations are given along with a method to recover selected seedlings and examples of use.

4.1 Introduction

Rice is one of the most important crops and is consumed by more than half of the world's population. Soil salinity is a major and increasing problem limiting rice growth and leads to huge yield losses every year. The search for new cultivars with improved tolerance to salt stress is a major goal in relieving this problem. This protocol gives an easy-to-follow procedure to select salt-tolerant rice lines for subsequent field testing.

4.2 Equipment

All equipment (tanks, trays, containers, drums and platforms) is dark coloured to minimise light penetration into the culture solution, thus reducing algal growth.

© International Atomic Energy Agency 2016

S. Bado et al., *Protocols for Pre-Field Screening of Mutants for Salt Tolerance in Rice, Wheat and Barley*, DOI 10.1007/978-3-319-26590-2_4

21

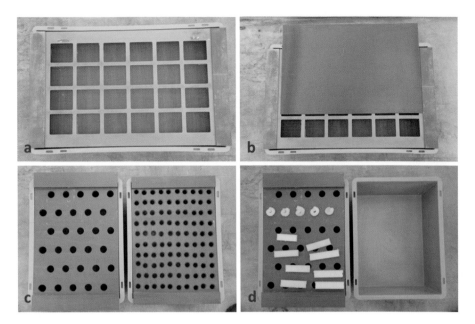

Fig. 4.1 Test and recovery tanks and plant platforms used in hydroponics

- Test tanks: These are made of plastic and have outside dimensions of $60 \times 40 \times 12$ cm and contain approximately 24 l each when full (Fig. 4.1). The size of tank can be changed to suit local conditions.
- Recovery tanks: These are made of plastic with outside measurements of $40 \times 30 \times 17$ cm. These hold approximately 20 l.
- Germination lids: PVC covers are used to blank out light; these sit over the PVC support platforms to provide darkness during germination (not obligatory). Germination lid dimensions: $50 \times 34 \times 2$ cm (Fig. 4.1b). The lids promote germination by helping to maintain humidity and temperature and cut out light.
- Support platforms:
  1. M_2 test platforms: PVC support platforms are made up with the dimensions $56 \times 36 \times 1.2$ cm to fit inside the top of a test tank. These platforms overlap the top of the test tank by 2 cm by gluing an additional sheet of PVC ($5 \times 36 \times 1.2$ cm) at both ends (Fig. 4.1a). M_2 screening platforms contain 24 rectangular compartments (6×7 cm) cut at regular intervals with a spacing of 1.2 cm. Each compartment can accommodate 100–200 seeds (useful for M_2 screening). Nylon mesh (fly netting) is cut to fit the PVC platforms (56×36 cm) and glued to the underside using PVC-V glue.
  2. M3 and other advanced generation/line test platforms: These PVC support platforms are made up with the dimensions $36.5 \times 26.5 \times 1.2$ cm. These overlap the test tanks by 2 cm by fitting an addition sheet of PVC ($5 \times 36 \times 1.2$ cm) at both ends (Fig. 4.1c). Round holes are drilled out

(100 round holes, 2 cm diameter). Two of these support platforms can sit on top of one big test tank.

3. Support platforms for recovery tanks: PVC platforms are made up with the dimensions $36.5 \times 26.5 \times 1.2$ cm. These overlap the tanks by 2 cm by fitting an addition sheet of PVC ($5 \times 36 \times 1.2$ cm) at both ends (Fig. 4.1c, d). Thirty equidistant open holes (2.2 cm diameter) are drilled into the support platforms (without mesh).

- Sponge strips ($10 \times 2 \times 1$ cm) (Fig. 4.1d).
- Storage containers for stock solutions: Stock solution can be prepared in small amounts and stored in the glasshouse or at room temperature for 1–2 months; mineral precipitation or the change in the covalence such as Fe or Cu in the solution is negligible over this period. The storage containers are air- and lighttight to allow long storage (1–2 months, Fig. 4.2). Nutrients for rice hydroponics have been described by Yoshida et al. (1976) and consist of six stock solutions (five for major elements and a sixth one for all microelements); for convenience, these are normally made up in 5 l amounts (Table 4.1).

Fig. 4.2 Six containers (5 l) for Yoshida stock solutions

Table 4.1 Constitution of stock solutions

Stock no.	Chemical	Amounts/5 l
1	NH_4NO_3	457 g
2	$NaH_2PO_4H_2O$	201.5 g
3	K_2SO_4	357 g
4	$CaCl_2$	443 g
5	$MgSO_4\ 7H_2O$	1 620 g
6	$MnCl_2\ 4H_2O$	7.5 g
	$(NH_4)_6\ Mo_7O_{24}\ 4H_2O$	0.37 g
	H_3BO_3	4.67 g
	$ZnSO_4\ 7H_2O$	0.175 g
	$CuSO_4\ 5H_2O$	0.155 g
	$FeCl_3\ 6H_2O$	38.5 g
	$C_6H_8O_7\ H_2O$	59.5 g
	1 M H_2SO_4	250 ml

- Storage drums for the working Yoshida solution: The working solution is made up using the six stock solutions and then diluted with distilled water in large drums. For convenience, the drum may be fitted with a submersible water pump to aid mixing, aeration and distribution into tanks. The solution may be prepared fresh or stored for incorporation in the next pH and volume adjustment (every 2 days). Large volumes of Yoshida solution (up to 120 l) may be stored in airtight and lighttight drums in the glasshouse for up to 1 week.
- pH meter.
- Electrical conductivity meter.

Note: Distilled water is preferred in making up Yoshida solution as local tap water may result in precipitation of minerals and will alter mineral concentrations that may affect salt sensitivity.

4.3   Plant Materials

Test materials should be compared against standard genotypes of known salt tolerance. Standards used at the Plant Breeding and Genetics Laboratory (PBGL), Seibersdorf, Austria, are as follows:

- Pokkali: Salt-tolerant wild type
- Nona Bokra: Salt-tolerant wild type
- Bicol: Moderately salt tolerant
- STDV: Moderately salt tolerant (induced mutant from IR29)
- Taipei 309: Salt susceptible
- IR29: Salt susceptible

The salt tolerance of the above standards in saline hydroponics has been correlated with the field performance (Gregorio et al. 1997; Afza et al. 1999). These standard materials can be requested free of charge from IRRI under a Standard Materials Transfer Agreement. Alternatively, local cultivars or breeding lines of known salt tolerance may be used as standards.

4.4   Setting Up Hydroponic Hardware

The screening is done in glasshouse conditions with day/night temperatures of 30/20 °C and relative humidity of at least 50 % during the day. The glasshouse should be disease free and well lit by natural or artificial lighting. The tanks may be placed on the floor or on the bench, but the surface should be as levelled as possible; tank water levels may also be adjusted using wedges.

Table 4.2 Composition of the working solution

Stock	Main element	Amounts of stock solutions needed for one (20 l) tank (ml)	Amounts of stock solutions needed for one (120 l) drum (ml)	Concentration of the elements in working solution (mg/l)
1	N	25	150	40.00
2	P	25	150	10.00
3	K	25	150	40.00
4	Ca	25	150	40.00
5	Mg	25	150	40.00
6	Mn	25	150	0.50
	Mo			0.05
	B			0.20
	Zn			0.01
	Cu			0.01
	Fe			2.00

4.5   Preparation of Hydroponic Solutions

The working solution is prepared as described by Yoshida et al. (1976) with adaptations made by Gregorio et al. (1997) (Table 4.2): each stock solution is shaken and 150 ml samples of each stock are mixed together and made up to 120 l. The pH of the working solution is adjusted in the drum to 5.0 with 1 N sodium hydroxide (NaOH) and 1 N hydrochloric acid (HCl) with continuous stirring (a pump may be used) to insure the solution is homogenised; this simultaneously aerates the solution.

4.6   Seed Storage and Seed Pregermination Treatments

Seed should be stored in dry, airtight containers at 4 °C. Germination of seed should be determined before testing as it is essential that seeds germinate uniformly (at the same rate) and that sufficient seedlings are available for testing. Some seed samples may have high seed dormancy; this can be broken by heat treatment at 40–50 °C for 2–5 days. Seed samples may also suffer from varying degrees of microbial contamination. This can be controlled by surface sterilisation by soaking in 0.8 % sodium hypochlorite (NaClO) for 20 min and then washing three times with water. Solutions of sodium hypochlorite can be easily made from commercial bleach (about 5 % NaClO). This treatment also promotes even germination.

4.7   Seedling Establishment in Hydroponics

The tests are normally conducted in a glasshouse set-up for rice: 30/20 °C day/night temperatures with 70 % humidity and 16 h photoperiod. Test tanks are filled with distilled water until the water level is about 1 mm above the mesh. The water level may be adjusted using wedges. Seeds are then placed into the wet compartments. For M_2 screening, 30–50 seeds from one panicle are placed into one compartment (6×7 cm) (Fig. 4.3a); for M_3 and advanced line testing, 5–10 seeds (or germinated seeds) are placed into each 2 cm diameter compartment; lines may be replicated within and among tanks (Fig. 4.3b). The test platforms are then covered with a lid for 1 week to promote germination in the dark. At day three, the water is replaced with half-strength Yoshida solution as vigorously growing seedlings will require nutrients. After 1 week, the platform of germinated seed is transferred to a test tank containing full-strength Yoshida solution to establish healthy seedlings prior to salt treatment. Seedlings are grown on to the two-leaf stage and should appear green and healthy prior to testing.

   Note: The test should not be carried out on unhealthy seedlings.

   Note: If seed samples are not clean and rotting occurs during germination, these must be removed. Seed may be surface sterilised prior to germination by soaking in

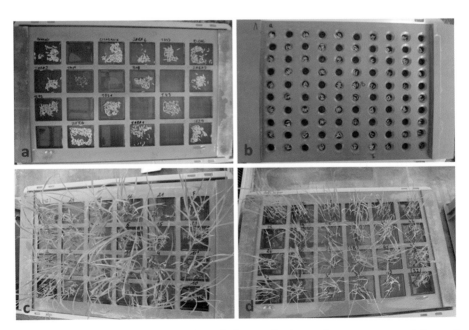

Fig. 4.3 (a) M_2 platform with rice seeds ready for germination; each compartment contains seed from one panicle per M_1 plant. **(b)** M_3/advanced line platform showing rice seed ready for germination; each compartment contains 3–5 seeds from each line. **(c)** Rice seedlings at the 2–3 leaf stage ready for salt screening. **(d)** Rice seedlings showing various degrees of salt injury

20 % Clorox solution for 20–30 min, followed by three rinses in distilled water. Clorox treatment also helps to promote germination.

4.8 Care of Plants in Hydroponics

Due to evaporation and transpiration, there will be loss of solution volume and pH change (algal growth may also contribute in pH fluctuation). Every 2 days (or thrice a week), the volume needs to be brought back to the level of full capacity (touching the netting in the platform compartments) and the pH adjusted to 5. Solutions can be changed by lifting off the platforms and placing them temporarily onto empty tanks and pouring the hydroponic solutions back into a drum where the bulked solution can be pH adjusted for the whole experiment in one step. Once adjusted, the solution is redistributed into the test tanks and the seedling platform returned. These operations also act to aerate the hydroponic solution. Alternatively, the pH can be adjusted on an individual tank basis, and more working solution may be added to make up the volume in each tank.

4.9 Salt Treatment

Salt treatment is applied at the 2–3 leaf stage, after 1–2 weeks of seedling establishment in full-strength Yoshida solution (depending on the rate of seedling establishment, Fig. 4.3c). The salt treatment is applied in one go and not incrementally. The test salt concentration is 10 dS/m (10 dS/m corresponds to 4.8 and 6.4 g of NaCl, respectively, in 1 l Yoshida solution and distilled water). Table 4.3 provides the conversion of NaCl in g/L in Yoshida solution for mmol and dS/m. Salinisation of the nutrient solution (working solution) is done for large volumes by adding dry NaCl in a drum containing Yoshida, dissolved and mixed using a submersible water pump. Salt is added until the 10 dS/m is reached; electrical conductivity is measured using an electrical conductivity meter (EC meter).

4.10 Scoring

Visual symptoms of salinity stress are reduced leaf area, whitish appearance of lower leaves, leaf tip death and leaf rolling. The technique for salinity screening is based on the ability of seedlings to grow in salinised nutrient solution. Standard genotypes are normally included in each test tank for comparison. Scoring is relative and carried out according to the standard evaluation system developed by IRRI with a score 1 for tolerant and 9 for sensitive (Table 4.4). Scoring is carried out at or around day 12 of salt treatment. At this stage, sensitive seedlings begin to die,

Table 4.3 Conversion table of NaCl in g/l added to Yoshida solution to mmol and dS/m

g/l	mmol	dS/m
0 (Yoshida solution)	0 (Yoshida solution)	1.17 (Yoshida solution)
0.42	7.19	2
0.94	16.08	3
1.22	20.88	4
1.76	30.12	5
2.56	43.81	6
3.1	53.05	7
3.66	62.63	8
4.22	72.21	9
4.78	81.79	10[a]
5.36	91.72	11[a]
5.92	101.30	12[a]
6.5	111.23	13
7.08	121.15	14
7.66	131.07	15[b]
8.26	141.34	16[b]
8.84	151.27	17[b]
9.46	161.88	18[b]
10.04	171.80	19[b]
10.96	187.54	20[b]
13.9	237.85	25
17.08	292.27	30

[a]Commonly used test concentrations for rice
[b]Commonly used test concentrations for barley and wheat

whereas intermediate genotypes show varying degrees of tolerance (Fig. 4.3d). Table 4.5 gives classification criteria for salt tolerance based on known standards.

Note: Scoring may be carried out at each day of treatment if quantitative data are required. Growth curves may be plotted to study responses over time. The biomass of seedlings may be recorded for this purpose using shoot/root/whole plant weight (fresh and dry), plant height and tillering that can be scored during the qualitative evaluation over time. However, scoring should be carried out for longer than 12 days of salt treatment as it is at this point that growth reduction of susceptible seedling becomes most apparent, whereas tolerant seedlings show some growth increase (but reduced compared to control seedlings).

At day 12 of salt treatment, tolerant standards (Pokkali and Nona Bokra) show slight damage with leaf tips becoming brown; moderately tolerant standards (Bicol and STDV) exhibit more leaf damage with dead older leaves and younger leaves being green only at their leaf bases; susceptible standards (IR29 and Taipei 309) are dead.

Table 4.4 Relative classification: scoring test genotypes/populations against known standards

	Two standard genotypes used Pokkali and IR29	Three standard genotypes used Pokkali, Bicol and IR29
Salt tolerance classes	I: More susceptible than IR29 II: Equally susceptible as IR29 III: Moderately tolerant IV: Tolerant	I: More susceptible than IR29 II: Equally susceptible as IR29 III: Less moderately tolerant than Bicol IV: Moderately tolerant comparable to Bicol V: Less tolerant than Pokkali VI: Tolerant

Table 4.5 Assessment scores of seedlings with respect to relative salt tolerance

Score	Visual observation	Relative tolerance
1	Normal growth; no leaf symptoms	Highly tolerant
3	Nearly normal growth; but occasional white leaf tips and rolled leaves	Tolerant
5	Growth severely retarded; most leaves rolled, few leaves elongate	Moderately tolerant
7	Complete cessation of growth; most leaves dry and some seedling death	Susceptible
9	Most seedling dead or dying	Highly susceptible

Susceptible lines will die at the same time as or before the sensitive standard IR29.

Moderately tolerant lines will respond in a similar manner to Bicol.

Tolerant lines can be selected when Bicol begins to die or has died; these may be removed to a recovery tank.

Symptoms on selected tolerant lines may be compared to Pokkali to estimate the degree of tolerance.

In cases where no standard lines are available, the following table may be used to assess tolerance in seedlings (Table 4.5). This table has been adapted from "Screening rice for salinity tolerance" (Gregorio et al. 1997).

4.11 Recovery of Salt-Tolerant Lines

Selected tolerant seedlings are teased out of the test tanks with care taken to keep roots intact. The base of the aerial part of each selected seedling is then gently wrapped in a sponge strip ($10 \times 2 \times 1$ cm) and the seedlings inserted into a recovery tank (Fig. 4.1d).

Selected seedlings can be grown to maturity in these tanks filled with Yoshida solution changed every 2 weeks.

Table 4.6 Classification of salt tolerance in 41 rice cultivars from Iran[a]

More susceptible than IR29	Susceptible, equivalent to IR29	Moderately tolerant equivalent to Bicol
Anbarbo, Hazar, Hashemi, Sadri, Domsefid, Mehr, Neda, Kadous Tarom Mahali, Daylaman, Hasan Sarai, Saleh, Sangeh tarom, Amol-3, Ghil-1, Drafk, Salari, Bejar, Nikjou, Pooya, Sahel, Shafagh, Fajer, Tabesh, Shirodi, Line-147, Line-145, Line-54, Line-29	Champabodar, Mazandaran, Shahpasand, Gharib, Hasan saraii atashgah, Dom Siah, Ghashangeh, Line-144	Neamat, Ghil-3, Binam, Ahlameytarom

[a]Data from training fellowship (Mr. Masoud Rahimi) in the IAEA Technical Cooperation Project IRA/04035 entitled "Developing salt-tolerant crops for sustainable food and feed production in saline lands (INT5147)"

Table 4.7 Classification of salt tolerance in 50 rice cultivars from Myanmar[a]

More susceptible than IR29	Susceptibility equivalent to IR29	Moderately tolerant
Thone Hanan Pwa, Ye Baw Yin, Ekare, Pa Chee Phyu, Mya Sein, Shwe Kyi Nyo, Mwe Swe, Maung Tin Yway, Shwe Ta Soke, Zein Yin	Pin To Sein, Shwe Dinga, Mine Gauk 1, Kauk Thwe Phyu, Pa Che Mwe Swe, Lin Baw Chaw, Rakhaing Thu Ma, Emata Ama Gyi, Hnan War Mee Gauk, Imma Ye Baw, Ye Baw Latt, Ekarin, Ban Gauk, Pa Din Thu Ma, Bom Ma De Wa, Nga Kywe, Sein Kamakyi, Nga Kywe Taung Pyan, Kha Yan Gyar, Nga Kywe Yin, Paw San Bay Kyar, Kamar Kyi Saw, Saba Net, Bay Kyar Gyi, Paw San Yin, Pathein Nyunt, Nga Kyein Thee, Mee Don Hmwe, Byat, Law Thaw Gyi, Moke Soe Ma Kywe Pye, Taung Hti	Aung Ze Ya, Ekarin Kwa, Ye Baw Sein, Gauk Ya, Nga Shink Thway, Paw San Hmwe, Saba Net Taung Pyan, Sit Pwa

[a]Data from fellowship training (Mr. Tet Htut Soe and Ms. Nacy Chi Win) in the IAEA Technical Cooperation Project MYA/06031 entitled "Human resource development and nuclear technology support"

4.12 Examples

The following tables summarise salinity data from results of seedling hydroponics screening carried out at the PBGL on materials from Iran (Table 4.6), Myanmar (Table 4.7) and Vietnam (Table 4.8).

Table 4.8 Summary of results after screening mutants of the cv. TAM (HNPD103, QLT4, T43, TDS4, TDS5, TDS1, TDS3, TL2 and HNPD101) and a Basmati rice mutant from Vietnam[a]

More susceptible than IR29	Equally susceptible as IR29	Moderately tolerant
TAM (parent)	QLT4	TDS1
HNPD103	T43	TDS3
BAS370 mutant	TDS4	TL2
	TDS5	HNPD101

[a]Data from training fellowship (Ms. Doan Pham Ngoc Nga) in the IAEA Technical Cooperation Project VIE/066011 entitled "Enhancement of quality and yield of rice mutants using nuclear and related techniques (VIE5015)"

References

Afza R, Zapata-arias FJ, Zwiletitsch F, Berthold G, Gregorio G (1999) Modification of a rapid screening method of rice mutants for NaCl tolerance using liquid culture. Mutat Breed Newsl 44:25–28

Gregorio GB, Senadhira D, Mendoza RT (1997) Screening rice for salinity tolerance. IRRI Discussion Paper series 22, vol 22. IRRI, Manila, p 30

Yoshida S, Forno DA, Cock JH, Gomez KA (1976) Laboratory manual for physiological studies of rice. IRRI, Las Banos, Laguna, p 83

Chapter 5
Protocol for Screening for Salt Tolerance in Barley and Wheat

Abstract A simple protocol is presented that tests salt tolerance in wheat and barley seedlings. The method is based on a glasshouse, aerated hydroponics test in which salt is added to the nutrient hydroponic solution in which the seedlings are grown. A list of equipment is provided including hydroponic hardware and stock solutions. Advice is given on seed storage prior to use and pregermination treatments that promote even germination of test samples. Salt treatments commence after seedling establishment in hydroponics at the 2–3-leaf stage. Visual symptoms of salinity stress include reduced leaf area, whitish appearance of lower leaves, leaf tip death, leaf rolling and seedling death. Recommended test salt concentrations for testing wheat and barley are given along with a method of recovering selected plants. Examples of protocol used are also given.

5.1 Introduction

The protocol for rice needs to be adapted for other cereals such as wheat and barley. Changes are required for germination, aeration of hydroponics and test concentration. Wheat and barley seeds do not germinate well when submerged and therefore cannot be germinated in the hydroponics platforms. Seeds are therefore pregerminated and young seedlings are transferred to hydroponics. Also, wheat and barley cannot tolerate anaerobic growing conditions, and their roots need to be aerated in hydroponics. Moreover, wheat and barley are more tolerant to salt than rice and therefore are tested at higher concentrations.

5.2 Adaptations of Rice Protocol to Wheat and Barley

5.2.1 Germination

Germination of wheat and barley may be improved by pretreatment with 0.8 % sodium hypochlorite; this serves to surface sterilise the seed and promotes more

© International Atomic Energy Agency 2016
S. Bado et al., *Protocols for Pre-Field Screening of Mutants for Salt Tolerance in Rice, Wheat and Barley*, DOI 10.1007/978-3-319-26590-2_5

synchronous germination. Solutions of sodium hypochlorite can be easily made from commercial bleach (about 5 % NaClO). The pretreatment involves soaking seed in 0.8 % NaClO for 20 min and then washing three times with water. Seed (up to 50) may be placed onto wetted filter paper in Petri dishes (4 ml water per 9 cm Petri dish, water should not cover the seed). The Petri dishes are then placed in the dark at 4 °C (fridge) for 48 h; this low temperature facilitates uniform germination. Seeds are best germinated in sandwich blots, which produce vertical root systems (Fig. 5.1). The pretreated seeds are placed into sandwich blots which are held in a rack placed in a dish of water at room temperature (17–25 °C) in lit conditions. After 5–6 days, seedlings of uniform development are removed and placed into hydroponics. Alternatively, pretreated seed in Petri dishes may be transferred to room temperatures (17–25 °C) in the light for another 48 h, replenishing water when necessary (when the Petri dish is tilted, there should be an excess of about 1 ml water). After 4 days in the light, germinated seeds should have the first leaf emerging from the coleoptile and 3–8 roots (Fig. 5.1b). These seedlings are removed individually and placed inside a sponge collar and inserted into the hydroponic system (Fig. 5.2a, b).

5.2.2 Hydroponic Solution

The hydroponic system is the same as that described for rice except full-strength Yoshida solution is used from the beginning of hydroponics. Hoagland's solution may also be used as an alternative (this is commonly used in research of wheat and barley) (Hoagland and Arnon 1950) (Table 5.1).

Fig. 5.1 (**a**) Sandwich blots for germinating wheat and barley seed to obtain vertical root systems. (**b**) Opened sandwich blot with barley (or wheat) seedlings; those of uniform growth are transferred to the hydroponic system

Fig. 5.2 (**a**) Wheat seedlings set into the hydroponic system in the glasshouse. (**b**) Wheat seedlings at the 2–3-leaf stage ready for salt screening treatment

Table 5.1 Composition of Hoagland's solution

Chemicals	mg l^{-1}
$NH_4H_2PO_4$	115
H_3BO_3	2.86
$Ca[NO_3]_2 \cdot 4H_2O$	945
$MgSO_4 \cdot 7H_2O$	250
$MnCl_2 \cdot 4H_2O$	1.81
KNO_3	607
$FeSO_4 \cdot 7H_2O$	5.0

Seedlings are grown on to the two-leaf stage and should appear green and healthy prior to testing.

5.2.3 Aeration

Wheat and barley and other cereals cannot be grown in hydroponics without aeration (rice can be grown without aeration because it has specialised aerenchymatous root cells). Aeration is supplied by pumping air into the hydroponics through plastic tubes feeding metal aeration rods. The tubes are attached to a central ring tube, and a ring system provides uniform pressure; feeder tubes are fitted to the central ring. The ends of the feeder tubes are fitted with 20–30 cm long steel tubes that are perforated with small holes (about 1 mm in diameter) to allow small air bubbles to escape (Fig. 5.3). Metal aeration tubes are used as they sink to the bottom of the tanks and do not interfere with root growth.

Some convenient sizes which we recommend are:

- Ring tube: internal diameter 25 mm
- Feeder tube: internal diameter 0.6 mm
- Aeration rod: internal diameter 0.6 mm

Fig. 5.3 Aeration system used for wheat and barley. (**a**) Steel tube perforated with small holes (aeration rod); the steel tube should be of sufficient weight to lie horizontally at the bottom of the hydroponic tank and not interfere with the root system of the seedlings. (**b**) Feeder tube with steel tube end piece for aeration placed into hydroponic tank and showing the gentle air bubbling. (**c**) Aeration system showing gentle air bubbling in the tanks. (**d**) Aeration system and support platforms ready for growing wheat and barley in hydroponics

5.2.4   Glasshouse Conditions

The test is normally carried out in a glasshouse set-up for temperate crops: 20/15 °C (day/night) with a 16 h photoperiod.

Aeration rods can be made from 0.6 mm steel tubing cut to length with one end sealed by clamping.

With this system, 6–10 hydroponic tanks can be accommodated on a bench of 2×1 m (Fig. 5.3c, d).

As soon as seedlings go into hydroponics, they are aerated.

5.2.5   Test Salt Concentrations

Plants differ greatly in their tolerance of salinity. Wheat and barley are more tolerant to salinity than rice. Therefore, different salt concentrations are used. Typical concentrations used for screening wheat and barley are 15–20 dS m^{-1} or 150–200 mmol NaCl, compared to 10 dS m^{-1} or 100 mmol NaCl for rice.

Salt treatments are applied to seedlings established in hydroponics (two-leaf stage in Hoagland's solution). This is done by transferring the platform of seedlings from control tanks to hydroponic tanks containing the test NaCl concentration in Yoshida solution. This is a one-step method. An alternative is to add salt in daily increments of 25 mmol until the test concentration is reached (this will reduce the effect of osmotic shock).

Reference

Hoagland DR, Arnon DI (1950) The water culture method for growing plants without soil. College Agriculture circular No. 347, College of Agriculture, University of California, Berkeley

Printing: Ten Brink, Meppel, The Netherlands
Binding: Ten Brink, Meppel, The Netherlands